OPEN CARDS

Tiny Bud

Published by New Generation Publishing in 2019

First Edition

ISBN: 978-1-78955-664-3

www.newgeneration-publishing.com

New Generation Publishing

OBSERVATIONS

Observation 1 – Medical Science Is Able To Clone A Human But Not A Person

Observation 2 – Intelligence Is Not Part Of A Person's Character

Observation 3 – The Human Brain Has No Thinking Functions

Observation 4 – Alcohol Affects The Brain, Not The Mind

Observation 5 – Alzheimer's Cannot Be Cured, Could Be Delayed

Observation 6 – The Human Life Averages Some Twenty Thousand Days

Observation 7 – There Is No Purpose of Life

Observation 8 – A Man Without Identity Is A Human

Observation 9 – Good News And Bad News Are Just Information

Observation 10 – Both Music And Noise Are Simply Sounds

Observation 11 – AI Lack Ability To Imagine

Observation 12 – Not Half Empty, Not Half Full. It's Half A Glass

Observation 13 – Science and Mathematics Have No Identity

Observation 40 – Without Mirror, You Can Never See Your Own Eyes

Observation 42 – Character Of A Person Is Not Found In His DNA

Observation 43 – Confidence Is Something That Is Developed, Not Practiced

Observation 44 – Space Is Infinite

Observation 45 – Gravitational Force Has Yet Been Completely Understood

Observation 50 – It Is Dangerous If A Blind Person Does Not Know He Cannot See

Observation 51 – We Board A Plane Because We Assume The Pilot Wishes To Stay Alive

Observation 52 – One Who Has Never Woken Up Does Not Know The Difference Between Being Awake and Asleep

Observation 54 – Nothing Is Bigger Than Nature

Observation 55 – We Are All Prisoners Of Our Own Minds

An Introduction

In his last published book after his passing, titled 'Brief Answers to the Big Questions', theoretical physicist and cosmologist Professor Stephen Hawking went as far as saying: 'We are each free to believe what we want, and it's my view that the simplest explanation is that there is no God.' (Pg 38)

These were the assertions made by Professor Hawking:

1. God is more a definition of the laws of nature rather than a sentient being.
2. The universe was created after the Big Bang, which is the exact moment of creation.
3. Positive energy created and neutralised by negative energy is stored in the universe.
4. Time only began at the moment of the Big Bang. There was no time before the Big Bang, hence no time for God to create the world and the universe.
5. Einstein's theory of general relativity breaks down near the Big Bang and hence it is called a classical theory.
6. There are certain laws that are always obeyed which could be referring to everything in the universe constantly being in a state of equilibrium.
7. He predicted the realisation of 'There is no God' to happen by the end of this century.
8. He wrote that although we are yet to have a complete picture, we may not be far off.
9. He believed that there is no heaven and no afterlife.

In the documentary television series titled Cosmos: A Spacetime Odyssey, astrophysicist Neil deGrasse Tyson said there is nothing recorded before the Big Bang. Nobody has any idea what existed prior to the Big Bang some 14 billion years ago. In Tyson's lecture titled 'The Inexplicable Universe: Unsolved Mysteries, Lecture 1 History Mysteries', he quoted 'of all the amazing things about the universe, we know so much about the universe but there is even more that we don't know'.

The search for the origin of life and the universe is never ending. The limitations of the human minds need a beginning and an end.

Numerous theories and speculations were presented as to how the universe came about including the speculation (in The Inexplicable Universe series) that, possibly some microbes from Mars landed on earth, seeding earth with early life. There was even speculation that all living things came from the same ancestor. Single cell theory speculates on this possibility. String theory was another attempt but there is no evidence by this theory. Dark matter problem is still the longest-standing problem in modern astrophysics and it persists to this day.

E equal MC squared, the theory of relativity does not come to total agreement with quantum physics. Two physicists named Professor Andy Albrecht (University California) and Joao Magueijo (Imperial College), in trying to understand the weaknesses in Einstein's formula, began to contemplate that there may be an infinite number of Big Bangs. The understanding that Big Bangs, in reference to time, is itself an infinite event could be helpful to understand the universes. The new view of the cosmos presented by these two physicists is the one in which the universe no longer has a beginning and an end but is eternal and an endless cycle of Big Bangs.

Einstein realised his formula had weaknesses. If his theory was right, the universe would have collapsed by now. Hence, he included a cosmological constant Lambda in his theory but subsequently removed it when astronomer Edwin Hubble managed to convince Einstein that the cosmological constant Lambda is not necessary to keep the universe in equilibrium against the force of gravity in 1932.

G = 8 Pi T being the original formula, Einstein adjusted it to G + Lambda g = 8 Pi T

This equation expresses the relationship between mass-energy and time space. It encompasses the structure of the entire universe. When Einstein used this equation to describe the entire universe, he ran into problems. The theory of relativity breaks down when gravity comes into the picture. This being highly imperative seeing that gravity is the dominant force in the universe. He lived with his unsatisfactory formula until his last day.

Stephen Hawking's answers were brief and the big questions remains. If the present physicists and medical scientists accept current knowledge as forever fixed, humans will forever struggle to find the answers about the origin of life and the universe.

Big Questions

1. Does God exist?
2. Why doesn't God answer everyone's prayers all the time?
3. Why can't we question God?
4. Why does God allow evil to exists?
5. How does the origin of life point to the existence of God?
6. Is Darwin's theory of evolution accurate?

There are many theories put forward but still as at now, the questions stated above remains. Talks by scientists and religious speakers continue. Questions from the audience keep coming. The chicken and egg conundrum always persist. The very fact that many scientists and physicists attend the talks by leading religious speakers and ask questions, shows that even they are still searching for the truth.

Even with the belief that there is no God, atheists do have other pertinent questions such as how did the universes, the earth, the planets, living things and non-living things come about? The physicists, atheists and religious leaders have yet to come to a complete agreement. One frequently-debated questions is 'why does God allow evil to exist?'. No satisfactory answers agreed by all parties have been given so far. Could it be that human minds, by its own nature, are more inclined to the pull of defiled thoughts? We will explore this in the latter sections.

Coordinates

Imagine a coordinate grid with x-axis and y-axis. One of the religions may assume the coordinate of (4, 5) as their version of origin. As far as this religion is concerned, after so many years of practice by its followers based on the origin coordinate of (4, 5), this religion believes that it is at the origin (0, 0) and the rest of the points are taking positions as coordinates measuring from this assumed origin (4, 5). Similarly, another religion may assume another coordinate as the origin. In this model of explanation, the origin is assumed to be (0, 0). It may well be argued that this model assumes the origin just as others have assumed the origin.

Atheism

The evolutionary biologist and outspoken atheist, Professor Richard Dawkins, has been saying that faith has been given a free ride for too long. The author, David Berlinski, is of the opinion that fossil records do not support the theory of evolution. British stand-up comedian, Ricky Gervais, when asked if he believes in God, responds he would rather believe in his dog. Australian stand-up comedian, Jim Jefferies, is similarly a well-known atheist. All atheists do not believe in the existence of God but have difficulty to prove their claim. In essence, it is just another belief system. At best, they can only say how can there be a God? It is their belief based on their intelligent reasoning but they are unable to elaborate especially when it comes to topics surrounding the moment of creation.

Going back to where we started, Stephen Hawking went as far as saying how can there be God. This is because, to his understanding, time does not exist before the moment of creation. This understanding that time only exists from the moment of a Big Bang is a wrong view from the very start. Another wrong view by Stephen Hawking is that he believed there is no afterlife.

Observation 14 - What Science Is Unable To Explain, We Call Mysteries

Every phenomenon and everything in the universe can and should be explained. What science has yet to be able to explain, we call it mystery. It is more of a keep-in-view attitude. If physicists considered present scientific knowledge as forever fixed, there will forever be unknown mysteries.

Observation 6 – The Human Life Averages Twenty Thousand Days

In most countries, human beings have an average life span of 80 years. By nature of measuring in years, we conveniently make ourselves believe there is plenty of time to go. With this, humans continue to submerge themselves in their daily routines and stressful lives.

A human life ends in the following scenarios:

1. When the energy in one's mind is no longer sufficient to support human life
2. When the biological strength of the human body expires
3. When all the conditions are present for a deadly accident to occur

Using a burning candle as an analogy:

1. The wit is used up before the wax is completely finished, similar to scenario 1 above
2. The flame dies off when the wax is completely finished, similar to scenario 2 above
3. A strong wind blows the flame away, similar to scenario 3 above

Observation 7 - There Is No Purpose Of Life

Being helpful, compassionate, kind, generous or religious is a way of life. Not the purpose of life. Mahatma Gandhi practiced a no-violence way of life. Mao Zedong led an authoritative way of life. Mother Teresa lived a compassionate way of life. The purpose of life is always referred to a person. If we assume there is afterlife, whereby the human body and mind only lasts an average of twenty thousand days, then a person can be reduced to a human being. Seeing that the mind continues to migrate from body to body, depending on the level of the mind's energy, a person only exists as human and body for a definite period. The existence of a person is only apparent truth, but the ultimate truth is that the person only existed as human.

Observation 8 - A Man Without An Identity Is A Human

A person with an identity is an apparent truth. Ultimately, he is a human. There are approximately 7.7 billion humans living on this earth. That is the ultimate truth. Ignorantly, humans created a world out of the planet earth by drawing boundaries, identifying themselves with nationalities, race and religions. Conflicts in the world arise due to a clash of identities. Ignorantly, many humans do not know that nature does not recognise identities. Identities that are derived from nationalities, race and religions only apply to apparent world but not the ultimate world.

Identities Are The Source Of Today's Problems In The World

In the song by John Lennon titled 'Imagine', the lyrics clearly showcase the problems caused by identities. Unnecessary deaths in the name of religion and ignorant killings due to different nationalities are all cause by a clash of identities.

The ultimate truth is that we are all citizens of the world and there should be only one race: the human race. To help a blind person cross the road is a wholesome act without any identity of religion or race or nationalities. As quoted by Professor Richard Dawkins, faith and belief in god has been given a free ride for too long. It can be harmful though only a handful of extremists carry out acts of terrorism. It is uncertain whether September 11 is an act by the US government when this conspiracy was put forward. It is uncertain when terrorists claimed responsibility and victory. What is certain is that 2 to 3 thousand human lives were lost in this terrorist act. The Holocaust is another such example.

Deep-Rooted Ignorance

Standing at the top of the tallest tower in the city, a tourist used a telescope to locate a specific building. If the tourist does not aim the telescope in the right direction, he would never locate the building no matter how long he spends. Having the right direction from the start of the effort to locate the building is important.

Similarly, in the effort to try to understand life and nature, if we start with the wrong viewing angle, we will always end up with lots of mysteries. To assume the human brain does the thinking functions is a wrong view; to think that there is no afterlife is a wrong view; the concept of God is a wrong view. The Greek philosophers have some right views but it was not complete. Plato contemplated on the concept of afterlife. The concept of Form was a close attempt to distinguish ultimate truth from apparent truth. To fully understand life and the universe, one has to have the complete right understanding of nature and the laws that govern it. In his understanding of the universe, Stephen Hawking is of the opinion that time only existed at Big Bang. This is another wrong view. Single cell theory is a wrong view. Evolution is a wrong view. Time only starting after Big Bang is a wrong view. No afterlife is a wrong view.

To stand a chance of having the complete right understanding of the origin of life and universe, one has to:

- Eliminate all of the wrong views
- Have an understanding of the phenomenon of infinity with regards to the abstract element (refer to Pg. 10, Before The Big Bang)
- Have an understanding of the infinity nature of time
- Have an understanding of the deep-rooted human ignorance and the laws of nature that governs it

Observation 3 – The Human Brain Has No Thinking Functions

The human brain has no thinking function. The mind does the thinking. The eye can only observe a colour's projective element. The ear can only observe a sound's projective element. The nose can only observe a smell's projective element. The tongue can only observe a taste's projective element. The reflective element of touch can only observe a touch's projective element. All the above projective elements are not motion. The five sensory organs being the eye, ear, nose, tongue and skin, therefore, cannot discern motion, but with their aids, the mind discerns motions of every kind. Every motion is discernible to the mind and the sensory organs merely assist the mind to discern them.

The process of thinking can be briefly explained as follows. When a thought occurs, the mind receives an impression in the form of an image of a situation. The impression enters the mind and remains fixed in it in the form of records. If the mind is powerful, it is possible for us to retrieve this impression retained in the mind. However, over time, with so many fresh impressions and old ones, the mind may not be sufficiently strong to retrieve the older impressions.

Due to the inadequacy of the mind's strength, it cannot penetrate through an overgrowth of impressions when thickly formed. The numerous layers of records grow as we age. To penetrate and retrieve recent records, the normal strength of a human mind is sufficient. Problems arise when the normal strength of a human brain deteriorates to an extent where it cannot perform the penetration and retrieval process even on recent records. The mind makes use of the sensory organs to do the thinking. The brain, which is biologically connected to the sensory organs, has the ability to control and direct the respective sensory organs to act accordingly.

When a person experiences something or is just thinking about that something, the reactions and reflections on all the sensory organs are exactly the same as the previous experience. However, the intensity is greater during the actual experience as compared to just when thinking about or recalling the incident itself. When all the similar reactions and reflections on the sensory organs are brought about with the help of the brain, the memory is then recalled.

When we think of an object, a faint reflection of the object springs in our inner eye. When we think of a sound we have heard, a faint reflection of the same sound springs in our inner ear. This applies to all our sensory organs. We think in terms of colours, sounds, smells, tastes, hot or cold, hardness or softness, passions and emotions. In each of these thoughts, we bring into activity its respective sensory organ. Whether we visualise a situation or whether actually experience it, there is an effect on the sensory organs, the difference being the degree of intensity. When we think of the taste of food for which we may have a particular partiality, our mouth secretes saliva. Thinking of sour plum being a good example. This is because the sensory organ of the tongue is activated when our thoughts are on the taste of a particular food.

Memories are stored in the mind just as movies are stored in a videotape. A remote control is used to perform the playback. Memories are not stored in the human brain just as records are not stored in the remote control. The human brain has only the same function as the remote control. All the mechanisms in a video player are what we know as the sensory organs. Without the remote control, the video player is still able to function, similar to one who is brain-dead. The mind will transmit instructions to the brain for the latter to perform the tasks, similar to someone who has to operate the remote control in order to rewind a movie on the video player. When the brain cannot operate or control the sensory organs to activate them and stimulate them to the similar or less intense situation as the actual experience, those memories consequently cannot be recalled. Without the rewind button on the remote control, the playback of the videotape is not possible. Four out of the five sensory organs (eyes, ears, nose, and tongue) are all fixed around the head while the fifth sensory organ (skin) is spread out across all regions of the body including the head. When we think, these five sensory organs are provoked, making us feel as if our thinking occurs in the head. Actually, the human brain neither thinks nor store records. It merely controls and coordinates the sensory organs.

When a doctor administers general anaesthesia, the metabolic rate of the patient's brain decreases tremendously. When the brain loses its ability to control and operate the sensory organs, it will ultimately result in amnesia. In the case of local anaesthesia, the brain is kept under normal conditions while the area of the nerve fibres is blocked by the binding effect of the local anaesthesia to the sodium channel. The impulse condition alongside the nerves are blocked, thus causing analgesia or loss of sensibility to pain.

When we look at an object, the reflective element present in our eyes reflects to the mind the projective colour element that is present in that object, and thus the mind receives a colour picture of the object before our eyes. Hence, seeing. In British illusionist Darren Brown's television series titled Fear and Faith, three audience members were put into an enclosure of complete darkness. After a while, one of them claimed that she saw a girl-like object or perhaps even a nun. What happened here is that the mind created the formation of the image and reflected in the mind even though there is absolutely nothing in that enclosure. As far as that participant is concerned, she saw something. This is somewhat similar to when someone says that they saw a ghost. As long as the image is depicted in the mind, the person will believe that they are seeing that image, whether it is through the sensory organ of the eyes or an image purely manufactured in the mind cause by overwhelming fear. In the same episode, Natalie, a stem cell scientist who is an atheist, was converted to a religious believer. Human minds are weak and constantly under influence such as being surrounded in an imposing environment like a congregation or in a worship premise. This is where human beliefs can be changed. Similarly, if our reflective element of the nose were to contact the projective element of smell, the mind takes an impression of the presence of smell. If the projective element of taste present in an object were to contact the reflective element of our tongue, that reflection is absorbed into the mind, thus forming the sense of taste. If an object emits the projective element of sound and that element were to get reflected by the reflective element of our ear and it is carried to the mind, the sense of hearing is then felt by the mind.

One of the functions of human mind is to maintain the human body's temperature. When the mind leaves the body, the dead body will submit itself to the outside temperature. Due to atmospheric conditions, no substances stay forever. A human being is a combination of a mind and body mechanism that exists for a limited duration of time. The human mind and body are two different entities that are strongly connected, with one constantly affecting the other, and vice versa. It is the conscious human mind with the help of a brain, controlling sensory organs and other human parts that enable the human to walk on two legs. When we form a large number of thoughts in our mind, these thoughts will gather necessary strength in order for us to do the action. The human mind does not extend outside our physical body. Imagine a white colour chalk being the non-living matter soaked in a blue solution. The mind is represented by the blue colour. The intensity of the mind is greatest at the human heart while least intense at the end of the hairs and the tip of the nails.

Observation 1 - Medical Sciences Is Able To Clone A Human But Not A Person

Cloning has been known to be present in the history of medical sciences for some time. In 2018, Jennifer Doudna introduced CRISPR-Cas9 where genome editing technology was possible which, if successful, would avoid unnecessary diseases. When we describe the character of a person, we are actually describing the formations in the mind of that person. However, all these formations are not found in the DNA. Hence, medical sciences can successfully clone a human with the same DNA but not a person with the same character. Intelligence can be found in the human DNA. Hence, science can clone another human with the same intelligence but science cannot make the cloned person to have the same character as the original human. By the explanation of this model, there is nothing morally wrong to have a human who has some diseases avoided in his life.

Observation 2 - Intelligence Is Not Part Of A Person's Character

When we describe the character of a person, we are in fact describing the mental formation of that person. A person can be confident, compassionate, greedy or egotistical. All these mental formations are not found in the DNA of a human being. Intelligence is.

Observation 17 – It Is Not Possible To Not Know Something That We Already Know

When we come to know something, it is recorded in our minds. These records cannot be erased. It might be a situation where the mind, even with aids of the sensory organs, is unable to retrieve the records. Alzheimer disease being one such case. When the memory arises in the mind, we can choose to neglect it but we cannot erase it neither can we ensure it is not in the mind.

Observation 19 - We Cannot Control The Brain But We Can Control The Mind

The brain is a biological organ that can be influenced by intoxicant materials or medications. We can only avoid the brain being affected by these factors which can cause the brain to be in an abnormal condition. Beyond that, in our daily lives, we cannot control the brain, we can only control our minds.

Observation 4 - Alcohol Affects The Brain, Not The Mind

The mind gives instructions to the brain. The brain needs to be under normal working conditions in order to perform the task as instructed. However, under an intoxicated condition, the brain will not be able to function normally. Depending on the extent of the intoxicated condition, the brain is able to communicate and function but not perfectly. A drunk driver wishes to walk straight as requested by the traffic police. The mind of the driver wishes to walk straight but because the brain cannot perform the task perfectly, hence the driver may not pass the test. Under normal circumstances, we can control our mind when our brain is in normal working condition. Medical scientists can operate on human brain, but the mind can never be operated on. Humans can walk straight on two legs because the mind is being aided by the sensory organs.

Observation 13 - Science and Mathematics Have No Identity

Science is a subject to discover and explain nature. Science can flourish because there is no identity. Similarly, mathematics is an education without an identity. There is no English mathematics or Indian science.

Observation 18 - Time Is Infinite

Time existed even before Big Bang. In fact, time exists in the form of infinity which has no beginning and no end. This contradicts Stephen Hawking, who in his last book, states that time only existed after Big Bang which meant that there was no time for God to create the universe. Time does not distinguish day and night. Present is the point separating the past and the future. Mathematics will collapse without infinity.

Before The Big Bang

Time existed before the Big Bang. Before the Big Bang, the entire nature is made up of four abstract elements (refer to Pg. 19, There Is No Origin Of Life):

1. Abstract element of earth
2. Abstract element of air
3. Abstract element of water
4. Abstract element of heat

These elements float throughout the entire universe which was in a state of complete darkness. These abstract elements continue to exist from the infinity past to the infinity future. They exist only in the form of energy.

Predominantly, the abstract elements of heat are floating in this state of blank darkness. After some time, they form themselves into physical structures of tiny atoms which float about in dark space. The physical systems then evolve to larger structures and, over a course of time, clouds of such atoms appear all over the space. During this period, the universes are still in a condition of blank darkness. These four elements are indestructible, because they are uncreated or unformed. Their existence will not end because their existence never began. These conditions are characteristics of all infinite phenomenon in the universe. These four abstract elements cannot be observed.

With regards to whether electrons can be observed, as described by Neil deGrasse Tyson in his series, electrons can never be observed, though recent scientific means have managed to capture the image of electron. In this model of explanation on the origin of life and the universe, it is assumed that these four abstract elements cannot be observed at the present time. These four abstract elements come together to form minds and non-living matters in the universe.

As the abstract elements evolve further, matters of living things and non-living things, planet and stars are formed. As for the minds, if the evolution in the mind is strong enough, it will gather other elements to exist as living things. While non- living things were formed without the presence of the minds. That is all there is in the universes. When the conditions are right, the formation of the universe, living things and non-living things, which were formed by these four abstract elements, will come together and start the entire universe. All data that we have ever collected on the history of life on earth such as fossils tell us that all lives began almost as soon as earthly conditions would allow it. This is the event commonly referred to as the Big Bang. It is the moment of creation. Stars and planets start to establish.

On every merger of the four abstract elements, subsidiary elements such as the projective elements of colour, taste and smell are formed. Complete substances discernible to our environment sense solidify.

New units of mind cannot be made, nor can these units of existing minds be destroyed. The units of mind merge with units of matter. The unit mind holds as many units of abstract elements as possible in the development of living things. The tendency of the unit mind is to gather into its hold as many units of abstract elements as it has the strength to control. When the limit is reached, it stops taking on additional units. When the unit mind takes units of the four abstract elements into its hold, their reaction to each other causes the formation of an atom.

The unit of mind begins to exist within an atom, having to process it to meet its requirements. The process of living things consists of a merger of one or more units of abstract elements with a unit of mind. The strength of the evolution current (refer to Pg. 20, There Is No Origin Of Life) in each of the unit mind determines which category of living things it will eventually be placed in once the accumulation process of abstract element is completed. If the evolution current in a particular unit of mind is adequately strong to sustain a human frame, it will settle and develop to become a human being. This process applies similarly to different species of animals, birds, insects and even bacteria.

During this time, only the beings developing directly on units of abstract heat are born without the support of parents. Conceptions did not take place to form living things. During this moment of creation, formation of living things does not need to come from conception. This contradicts with Charles Darwin's theory of evolution. David Berlinski's thorough research on evolution and his book 'Devil's Delusion' sheds some light on this topic. Briefly, David Berlinski says that fossil records does not support the theory of evolution. Depending on the strength of the evolution current, each living thing produces material substances and life elements in its physical system.

Imagine having one sack of rice and pouring it into a rice pot, which is a normal practice in the east for safekeeping and daily use. If the pot of rice is not consumed for a long time, black insects manifest within the rice pot. When the conditions are right such as when the surrounding temperature is acceptable and when the atmospheric pressure is suitable, the black insects begin to appear. The rice is a non-living object while the black insects are living things. If we equate the entire rice pot to the universe while the black insects represent all animals and human beings, we have a situation similar to the moments of creation. When all the rice is poured into the rice pot, all the potentials are there in the rice pot. When the conditions are right, the black insects, which have minds, will start to form. The black insects are formed based on the strength of the evolution of the insect mind.

The higher category of living things settling into material existence do not begin the formation from units of non-living matter. It begins in a unit of abstract heat before starting to acquire and retain units. If the minds are settling into non-parental conception, they begin to develop into various living elements by their own strength. The unit of mind thus brings together the ingredients necessary for its first atom. However, after the Big Bang, almost all living things need parental support. Due to adverse weather and lack of food, some species of animals such as dinosaurs became extinct. While beings are settling into non-parental conception, they begin to develop into various living things by their own strength. Trees and plants which have lives but unconscious minds start to form simultaneously. It should be noted that there is no higher category of living things discovered and described in the present world. Present science has never discovered new humans, elephants, tigers or lions born into this world without parental conception. However, for the lower category of living things, new species are formed and discovered. Science has identified two million species of plants, animals, and microbes on earth, but scientists estimate that there are millions more left to discover. The most-commonly discovered new species are typically insects, one category of animal with a high degree of biodiversity. Newly-discovered species of mammal are rare but they do occur typically in remote places that haven't been well studied previously.

Some animals are labelled as a new species only when scientists peer into their genetic code because they look externally similar to an existing species. These are called cryptic species. Some newfound species come from museum collections that have not been previously combed through and from fossils as well. 0n the 13th of February 2019, the BBC reported about a new mysterious frog species that was discovered in the Western Ghats of India. There are new forms of bacteria discovered every so often but this could be due to the mutation process of existing bacteria. In summary, during the Big Bang, all living things develop from the abstract element of heat. This applies to all category of living things. Living things formed during this time has no conception from parental support. After Big Bang, this phenomenon only happens to lower category of living things but not beings within the higher category such as humans, dinosaurs and animals.

The mind is a force. The force of the mind of a human being is greater than the aggregate force of trillions of units of mind of a bacteria. We have settled on this earth at a certain equilibrium – an equilibrium between a definite measure of mental and material forces with the other planets in our universe. The earth does not allow the material force of the earth to increase or decrease beyond the limits of its settled equilibrium. Minds are not subject to the control of the gravitational force of the earth. The occurrence of the physical gravitational system such as the gravitational fields of planets is dependent on the variations of the mental gravitational system developed in the mental system of each of the living beings in the universe.

The Big Bang happened approximately 14 billion years ago. Humans are still struggling to find a definitive answer as to the existence of God. Physicists are unable to completely understand the universe. One of the possible missing links could rest on the fact that all the living beings (humans, animals, insects, plants, vegetables, bacteria and all the other beings) are contributing towards the sustenance of gravitational force of this earth. If all the beings existing on this earth disappear from it, there would be no gravitational force. The formula, equations and model to explain the formation and the equilibrium established in the universe have yet to allow physicists to completely understand the universe. The relationship between the minds and the gravitational force, which is the least understood force of nature, is extremely important to completely explain the formation and equilibrium of the present universe. The weaker force of gravity compared to the electromagnetic force and the other two nuclear forces, fundamentally causes the discrepancy to completely describe the universe. If the gravitational force were to be slightly stronger, the earth, as we know it, will cease to exist. String theory, M-theory and Einstein's theory of general relativity are some of the unsuccessful attempts to explain the universe.

Another missing link could be the existing knowledge about the afterlife. This contradicts with Stephen Hawking who believed there is no afterlife. Human and other higher categories of living things formed during the Big Bang do not require parental support. The mind circulates from life to life. It cannot remain in any formations of substances perpetually. The reason is that every material formation is a beginning and every beginning will inevitably lead to an ending. The mind, which settles into a material human life, therefore, cannot remain in the same structure perpetually. It must, sooner or later, abandon it. After the Big Bang, humans were born through conception. During conception, the mother provides the new being with the life elements, food and shelter. Neither the mother, father, nor both of them together can provide the unit of mind essential for the growth of new being. The unit of mind that settles in the field of conception, appears from a foreign source, having migrated invariably from a previous existence.

The human mind does not extend outside our physical body. The forces of our mind, while remaining within our bodily self, constantly engage in the different tasks of maintaining the physical and sensual self. Its energy is not sufficient for it to be extended to focus on our past lives. However, at the end of the present life, the force becomes adequate to extend the vision outside in order to locate its next category of existence.

Once the migration of the unit of mind to the new category of living thing is accomplished, the vision of death ebbs. The forces of the mind now turn all their energy towards the formation of the new material system. During the course of the first beat of the mind (refer to Pg. 20, There Is No Origin Of Life) following conception, the first atom of the material body becomes completely assembled and solidified. The unit of living thing element settling in active strength begins to perform functions, in particular the food element and the heat element. The formation of the first atom, heart, eye, nose, tongue, and touch begin to form in approximately 70 days. The heart is the first organ to be developed. Abstract element of heat helps to construct the physical frame until youth but it also helps to decay the physical system at old age.

Every human life is an average of twenty thousand days. The death of a human being involves the separation of its unit of mind from its material body. In the case of normal death due to old age, the human will lose successively his or her functions of eye, ear, nose, tongue, and touch. The last organ to go out of action is the heart and that occurs only at the time of death. At the moment of death, all the forces of the unit mind, which are spread out across the material system, concentrate in the heart and a strong wave of vision flashes out. It is the final thought occurring in each material system and its confined to the last beat of the mind.

Depending on the strength of the mind, the exit vision rests on a new appropriate category of living things and in the meantime, the unit mind abandons its grip on its old material system. At the end of the cycle of the beat of mind developing the exit vision, the unit of mind migrates completely to the field where the exit vision is rested and there, it begins to develop a fresh physical system. Thus, the old system dies and the new system is born. However, for the unit mind itself, the only change amounts to leaving an old home and occupying a new one. This is afterlife.

It is due to some of these missing links that ignorance is evident amongst human beings. The ignorance causes pain and suffering in human life. Unnecessary deaths are caused by conflicts between countries and disagreement amongst religions, within the same religion, and even suicidal deaths. Due to ignorance and defiled mental factors, humans created this world out of planet earth. We humans mark boundaries to establish different countries and races, create the concept of God which in turn causes religious conflicts. The song by John Lennon titled Imagine where he says 'if humans can live live without countries and religions, there will be peace amongst the human race' rings true. Ignorance caused humans to create a conflict between Israel and Palestine. In actual fact, nature recognises only one race - the human race. Nature recognises the world which was created from the earth by humans as simply planet earth. There are 7 billion of humans of the same race living on this earth.

After The Big Bang

In our daily life, in order to keep ourselves physically healthy, we exercise. To keep the mind healthy, we must keep the mind still. In this aspect, to have good mental health refers to keeping sufficient positive deposits in the evolution current in the mind. The reason to keep sufficient positive deposits in the evolution current is to provide a condition where unpleasant situations do not arise or at least to postpone them. Providing a condition for the postponement or reduced scale of unpleasant happenings while at the same time inviting the pleasant to happen is an advisable practice. The Law of Conditions, which is one of the laws of the nature, together with the other two prominent laws of nature, the Law of process and the Law of Cause and Effect, should be well understood in order to make sense out of life. The Law of Conditions state that when all the required right conditions are there for something to happen, it will happen. It is a wrong view to say that things happen for a reason. Things happen for a reason often refers to the fact that since this has happened, something is going to happen in the future due to this happening. Whereas, when the Law of Conditions refers to this has happened, it is because all the conditions needed for this to happen are present therefore it happens. The Law of process, reiterates that everything is in the form of process which has a beginning and an end, and the Law of Cause and Effect is, in itself, quite self-explanatory. These laws of nature are constantly enforced.

Let's assume our minds are a mixture of crude and transparent oil. By nature, the heavy and transparent oil representing positive thoughts will reside at the bottom while light crude oil springs to the surface. The 7 common currents (refer to Pg. 19, There Is No Origin Of Life), evolution current and 6 other currents cause the formation of the 6 supplement currents. Therefore, together with this, the 13 forces of currents cause the formation of positive and negative mental factors. The strength of the evolution current in the mind which depends on the positive and negative deposits varies from moment to moment due to treatment or the reactions to the thoughts. The impression of the positive and negative thoughts goes into deposit in the evolution current as a potential and they will begin to grow.

After a period of time of accumulation, crowded growths of positive potential, upon reaching maturity, bears pleasant fruit. There are also cases of an abundance of positive potential so crowded that it can overtake negative potential and reach maturity to bear pleasant situations. The evolution current is centred in the heart. Impressions casted in the evolution current is in the form of sound, taste, touch and sensation. The impression can be in the form of emotion, love or hatred. Those impressions entering the evolution is a result of thoughts, positive or negative actions. They enter as potential and after they fructify, they remain in the evolution current. One of the laws of nature is such that when a human mind is going through suffering, the mind burns out the negative potential that has been accumulated in the past. By virtue of depleting part of the negative potential, the positive potential has the opportunity to surface. Pain and pleasures are the reactions. Action is positive and reaction is passive. In our life, passive thoughts are in abundance hence we tend to react more than taking action, when it comes to situations. Appearance of negative thoughts reduce present and future pleasures.

As mentioned previously, subsequent to the Big Bang, almost all living things require parental support to survive. Hence, there are unfortunate species of animals that succumb to extinction due to harsh weather, lack of food or destruction by humans or nature. The unit of mind holds as many units of abstract elements as possible in the development of material life. A unit mind is made up of 4 abstract elements but with multiple units of abstract element. The unit mind tends to gather into its hold as many units of abstract elements as it has the strength to control. When the limit is reached, it ceases to take on additional units. The force of mind cannot be added by adding two minds. Hence, when human minds are so negative, the strength is greatly reduced. Units of mind do not merge with other units of mind.

Only the 4 abstract elements and the element of sky are infinite. They are uncreated and unformed. The other 23 elements (refer to Pg. 19, There Is No Origin Of Life) are not and they are the subsidiary element of these 4 great abstract elements. Hence the 23 subsidiary elements have a beginning and an end. The mind has no beginnings and endings, cannot be made or destroyed. There are no new units of minds forming in universe. The 4 abstract elements circulate from substance to substance while the mind circulates from life form to life form. The strength of the 4 abstract elements are always the same but the strength of mind varies. Each unit of abstract element has a definite degree of force. When a substance is destroyed, the abstract elements migrate to other substances, or if there are no other substances available to merge, these abstract elements will exist independently in the form of energy.

When each material universe collapses, the abstract elements remain in the form of free energy distributed evenly over space. Each of us is composed of a unit of infinite mind and units of infinite abstract elements of matters. There are no fresh abstract elements in the universe. The strength of a substance can be increased by the proportion but not the strength of the 4 abstract elements.

If the force of a unit of mind seeking new accommodation is extremely low, the death vision will lead the mind into the abstract element of heat. When this happens, the mind is no longer a unit mind but an abstract element of heat. It's in a state of extreme suffering, otherwise known as hell. A subsequent unit of mind, which is more forceful, takes over the unit of conditioned abstract heat element and further evolves that unit of abstract heat element.

After a long period, that unit of abstract heat element is processed by many unit minds and finally, one unit mind which is stronger has enough strength to occupy conditioned element of abstract heat. When this unit mind gets stronger, it attracts unit of abstract units of water element, abstract units of air element, and abstract unit of earth element. The reaction of these units of abstract elements causes the formation of an atom. Then, the subsidiary abstract elements appear. The projective element of colour, sound, smell, taste and touch and the three feature elements of birth, existence and death appear simultaneously. The sky element appears at the same time to bind and preserve the individuality of each different unit.

When this unit mind completes its span of existence, some other unit mind with sufficient strength to develop life in physical structure of equal evolution will take over. As mentioned earlier, even amongst the most elementary of lives, only beings developed directly from units of abstract heat are born without the support of parents. The unit of mind with sufficient strength in their evolution current to develop into beings of the size of one or more atoms, have also the force necessary to process the other life elements in their physical system. Upon solidification of the atoms in their physical system, all life elements appear.

Owing to the fact that the speed of mind is extremely fast, living things appear simultaneously and so do all the non-living matters. As previously mentioned, all data collected on the fossil history of life on earth tell us that all lives began almost as soon as earthly conditions would allow it. Subsequent to the Big Bang, the offspring of an elementary being only needs a unit of the evolution element from the parents while for the rest of the units of element, it absorbs from the substance in the environment.
The order in which the various units of abstract elements are assembled, and the patterns into which they are woven together in forming the new structure, depends on the forces of the evolution elements acquired from the parental system. Owing to the forces of

the evolution element, the new atom formed will assume the same shape and other characteristics as the material system of the living thing from which the evolution element was derived. This process goes on from stage to stage, producing bigger and better material systems at each step.

As the evolution current of a unit of mind becomes stronger, the material system becomes more developed. More units of abstract elements enter it and also better life elements will be produced. As more units of abstract elements enter a physical system, the mass of the physical system also increases and subsequently, the physical system keeps increasing its mass and strength. When the physical system attains certain strength, the forces of sex set in and the process of evolution completely changes. In elementary life, both sex elements exist in active force. When the living things are more developed, the strength to continue the development of both sexes is insufficient. Hence, one sex is more dominant while the other sex is more dormant. The sex process in the higher category of living things causes severe repercussions all over the physical system and its reaction is spread to all organs.

In the higher category of living things, the female provides the evolution elements in the necessary merged form by acquisition and merging of a unit of evolution element from the male. Females provide appreciable assistance to the new beings taking conception by way of supplying food and shelter.

Life of the lower category of living things attached to abstract heat element do not need the food element because there is no strength to grasp more units of element. Higher category of living things requires the food element. The abstract element of heat, once settled down and if the mind is forceful, develops the food element. This food element will extract additional units of the elements until the physical system develops into an atom. Continuing this way, all the units of the 4 abstract elements merge together in the formation of a physical structure.

The higher category of living things retain large proportion of forces to produce intelligent sensations. Conversely, trees and plants, who have fewer sensations, uses their forces to build a large physical system. The unit of mind existing in us prevents the matter within our physical system from spreading out or allowing outside matter from entering. The atmosphere has the tendency to remove elements from our physical system. Food elements help to filter and take in fresh elements.

During the course of the first beat of the mind following conception, the first atom of the material body becomes completely assembled and solidified. The unit of life element settling in active strength begins to perform functions, particularly the food and heat elements. The formation of the first atom, heart, eye, ear, nose, tongue and touch begin to form in 70 days. The heart is the first organ to be developed. Circulation of abstract elements take place where the abstract elements try to escape but the mind takes in more to balance. The mind will always reserve some strength to develop various sensory organs.

During the stage of complete darkness before the Big Bang, each unit mind grasps more units of elements matter and thus causes concentration at some places and vacuum at others. Each unit of mind develops gravitational force to keep in concentration a measure of units of abstract elements. This could be the reason for the formation of Black Holes in relation to Stephen Hawking's analogy of making a hill by causing a cavity.

Every living thing in the material universe which exists contributes a measure of gravitational force to the universe. All the living things contributes toward the sustenance of gravitational force of this earth. If all the living things were to not be on this earth, there will be no gravitational force on earth.

Assuming there is no water from rainfall for a considerable amount of time due to insufficient gravitational force, the inevitable heat combusts the earth. In a distant future, when the combined strength of the evolution of minds in the universe is insufficient to hold planets and stars, the equilibrium is disturbed. As the combined force decreases, quantity of units of abstract matter also decreases. In other words, the units of mind lose their gravitational force to control a sufficiency of units of abstract elements and consequently the planets and the stars lose their strength to keep together the various substances in them. More of the substances from the planets and stars will radiate into space and disappear, and the conditions of the planets will become increasingly difficult.

The harsh climate followed by a scarcity of food makes human beings in the universe increasingly violent and cruel. The increase of cruelty amongst the beings in the universe reduces further their strength of the evolution current of minds. This causes conditions on the planets and stars to deteriorate further. The equilibrium of the universe is thus disturbed. When the earth loses the sustenance of gravitational force, the heat will become so immense that the earth explodes. By virtue of the fact that the universe is so vast, due to space being infinite, the explosion of the earth should not have as great as an impact that would cause the entire universe to collapse. Taking a large piece of tempered glass as an example, it can be strong and would be able to withstand a reasonable amount of force. However, there is a weak point in the entire piece of tempered glass and when this weak point is disturbed, the entire piece of tempered glass would collapse.

All matter crashes and all living things by this time are reduced in strength of their evolution current to an extremely low category of existence. They continue their existence by taking conception in units of free abstract heat. Free abstract heat does not produce a mass and therefore the living things existing in them are material beings without mass. Existence in the universe of free abstract heat is the most acute form of suffering in the universe. During the period of complete darkness, all minds are reduced to the free abstract element of heat. As mentioned earlier, it is an acute form of suffering commonly known as hell. There is no total charge of negative potential in the evolution current, and there is always a tiny tincture of negative potential. Except for some extreme cases, it is extremely difficult to remove the last tincture of negative deposit because the small quantity of negative deposit remains, serving as a binding agent to bind beings to material life.

As mentioned earlier, suffering reduces the forces of negative potential, and the beings remaining in such conditions liquidate most of their negative potential. After a considerable lapse of time, these beings liquidate their negative potential to a point that remnants of impressions of positive potential remaining in the evolution current of their mind increases. Thus, many units of minds acquire a sufficiency of evolution strength to cover their suffering physical systems of free abstract heat with a few units of the other abstract elements. When such strength is gained by these beings, they form themselves into physical structures of tiny atoms which remain floating about in space, in the state of darkness. As that strength increases further and further, the physical systems they evolve into become larger and in the course of time, clouds of such material beings appear all over space. At this stage, the activity of these tiny beings disturbs the homogeneous equilibrium of the then material universe. As the beings evolve further, planets and stars begin to form, living things and non-living things begin to appear. Owing to the speed of beats of abstract elements of living things and non-living things matters, the moment of creations take place simultaneously and once again the material universe is established. Beginning in this manner, the new universe grows, matures, decays, dies and ends again in a catastrophic dissolution, and the process goes on infinitely. Time is infinite. Time existed before the Big Bang. Time has always existed, exists now and will always exists. The number of Big Bangs is infinite. The earth and planets are in the habit of dissolving themselves in 100 trillion years. Since some 70 trillion years has passed since the last Big Bang, we are left with 30 trillion years before the next.

Observation 19 - There Is No Origin Of Life

A full and correct comprehension of the exact features of living things and non-living things is essential for the understanding of life, time, the universes and to determine whether there is an origin of life.

Living things have a presence of mind, non-living things do not. However, both living things and non-living things are made up of the 4 abstract elements (earth, air, water and heat).

For the non-living things, substances which are without the presence of mind, they consist of 13 elements, which are:

- 4 abstract elements of earth, air, water, and heat
- 5 projective elements of colour, taste, sound, smell and touch

- 3 feature elements of forces of birth, existence, and death
- 1 sky element.

For the living things, with the presence of minds, it has 28 elements.

The additional 15 living things elements, aside from the 13 stated above, are:

- 5 reflective elements of eye, nose, ear, tongue and touch
- 1 reflective element of the heart
- 2 sex elements
- 1 evolution element
- 1 food element
- 2 motion elements (physical motion element and sound motion element)
- 3 condition elements (element of release, inaction, adoptability)

The unit mind produces and carries 15 life elements. Life elements are the subsidiary of the 4 abstract elements. The mind beats at approximately 1/3 thousand trillion times of a second while elements in non-living things matter beat at approximately 1/176 trillion times of a second. Speed of repetition of elements in non-living things are 17 times slower than the speed of repetition of the mind. Inherent permanently in each unit of mind are 7 major currents, which cause the continuous flow of the 17 waves in each beat of mind. They are:

- current of survey
- current of condition
- current of distinction
- current of intention
- current of decision
- current of evolution
- current of recollection

Of the 17 waves, the first 3 waves belong to dawn season waves, followed by 12 existence waves and the last 2 are close season waves. The duration of each wave is 1/51 thousand trillionth of a second since the mind beats 3 thousand trillion times in approximately 1 second. This is the shortest duration measurable and is used as a multiple for any other measure of time.

- When the mind beats once, it has 3 seasons within each of the beats.
- Each beat of the mind has 3 mind moments.
- 1 beat of abstract element = 17 beats of the mind (51 mind moments)
- The abstract element takes 1 mind moment for birth, 49 mind moments to exist, 1 mind moment to die
- 17 beats of mind equal to 1 beat of non-living things matter

Particular attention is given to the current of evolution. The current of evolution has principal functions such as to ensure the continuance of the repetition process of the mind, to act as the repository of the mind and to maintain the release of the forces of the mind at the correct strength. Atom exists by elements in repetitive motion 176 trillion times in approximately one second. Non-living things are made up of the elements of earth, air, water, fire. Fire has an abundance of abstract element of heat and other elements but in smaller proportions. No substance or matter, living or non-living, is formed in the universe without the mixing of all the four principal elements.

In an atom, one of the four great abstract elements is reduced to unity. The smallest atom consists of only one unit each of the elements. During the splitting of an atom, when one of the elements is released from the grip, the substance ceases to exists and it disappears. Then, the substance assumes another form, which is not discernible to the sensory organs except for the mind. The atom, when dissected, assumes the form of a force. Each of the units of abstract elements carries 5 different currents of forces namely colour, smell, taste, sound and touch. These 5 subsidiary elements are substances. Each of the abstract elements carry in them 3 internal currents namely current of birth, current of existence and the current of death. The 4 principal abstract elements are in the form of energy. The entire universe is in motion. Mind is a motion. Sensory organs aid the mind. Thinking is a function of sensory organs. Without sensory organs, the mind does not think. Any object that moves against the law of motion of non-living things has life present within the sphere of its influence.

Wherever there exists a unit of mind, there exists life. The center of gravity of the mind does shift temporarily depending on the impact on situations and on other sensory organs. Eventually, it resides at the heart. We think in terms of reflection. Penetration of the exposure of the mind causes the mind to develop a glow which radiates into the mind causing a reaction that could either be of a positive or negative nature. Some of the positive mental formations are as follows:

- Confidence
- Mindfulness
- Shamefulness towards evil
- Fearfulness towards evil
- Loving charity
- Loving sympathy
- Equanimity
- Tranquilisation of actions of body
- Tranquilisation of actions of mind
- Lightness of mental properties
- Lightness of the mind
- Softness of the mental properties
- Softness of the mind
- Fitness of mental properties
- Fitness of mind
- Proficiency of mental properties
- Proficiency of mind
- Physical uprightness
- Mental uprightness
- Correct speech
- Correct industry
- Correct livelihood
- Universal sympathy towards all beings
- Universal desire for others' well being
- Knowledge

The negative mental formations include:

- Ignorance
- Shamelessness towards evil
- Fearlessness towards evil
- Turbulence

- Greed
- Wrong perception
- Conceit
- Hatred
- Jealousy
- Stinginess
- Repentance
- Sickliness of body
- Sickliness of mind
- Indecision

The entire nature is made up of abstract elements. Bacteria, insects, marine life, birds, animals and humans are the categories of living things in ascending order of their strength of the evolution current. If the strength of the evolution current is so weak to the point that it cannot even hold the frame of a bacteria and it is so thin that no present technological equipment is able to detect it, it will fall into a category commonly known as ghosts. Ghosts have extremely thin frames. Almost all civilisations around the world touch on the existence of ghosts or spirits but there is no hard evidence of them.

We can only suspect the presence of ghosts when within the sphere of the weak unit mind, there are things that can move against the law of physics. At the other extreme of the category, above human beings, are the happy spirits. These spirits have sufficient evolution current not to fall into the category of humans. If the evolution current is stored with an abundance of positive deposits, the mind may settle in a category higher and higher in the category of spirits. Again, since the formation of ghosts and happy spirits has a beginning, they are not perpetual. They will submit themselves to the law of process, which is one of the laws of nature. They have a beginning and an end.

Presence Of Mind In All Living Things

Sleep is a state of sub-consciousness, which is different from the consciousness present in plants and vegetables. The minds in plants and vegetables are unconscious minds. Plants and vegetables have minds, which is why they can grow and are considered living things. Anything that has a mind will grow whether it is a conscious mind or unconscious mind. All these minds contribute to the sustenance of the gravitational force. By hierarchy, the lowest is the category of ghost and on the other extreme end, is the category of happy spirits, which some civilisations refer to as deities.

We have dealt with the basic units of existence in the universe, the units of matter and the mind. Now, we need to study in detail how these units weave together in the formation of intricate webs of the material universe. The four abstract elements can never be made, nor can they be destroyed. We may be able to create a material object and destroy it, but no force in the universe can produce fresh abstract elements nor can any force in the universe destroy them. It is necessary to comprehend this fundamental truth.

There are two processes by which material objects in the universe come into existence: they are either made by man or animals or they form by themselves. The formation of each unit of element matter thus causes concentration at places and vacuum at others. In the words of Stephen Hawking, making a hill will cause a cavity at another place. All that is made can be destroyed. If not destroyed by man or animals, its destruction will come due to natural causes in time. Our material system, the body, was formed in stages. Similarly, its decay will happen throughout stages as well.

That we have been born into this material state of life is the reason that we must, sooner or later, die out of this material state. Each beginning is invariably linked to an end. Man begins by birth; therefore, he ends by death. The earth and the universe have a beginning and, therefore, they must also have an ending. Whatever has a beginning, be it a material object, a material system of a being, or a material universe, is always linked to an ending, and no amount of ingenuity by any being can alter this law of existence in the universe.

There are certain things in the universe which have no beginning which has no ending as well. These things never came to be; therefore, they will never cease to be. These things are the infinite factors of the universe and amongst them are the four great abstract elements. These abstract elements have never been made and can never be destroyed. Subject to their continuous process in the cycle of repetition, the four great abstract elements exist throughout all time. They have existed throughout the infinity of the past; they are existing throughout the present time and they will keep on existing throughout the infinity of the future time. These elements are indestructible, because they are uncreated or unformed. These conditions are characteristic of all infinite phenomena in the universe.

This infinite phenomenon forms a major part of the explanation of the origin of life and the universe. We must understand that the destruction or dissolution of material objects and substances does not cause the destruction of the abstract elements constituting them. The destruction of a material object or substance merely means the disintegration of the various abstract elements which combined to form the material object or substance. When, for instance, a substance like water is destroyed, the abstract elements which constitute the water migrate into other substances. If there are no substances available for these elements to merge with, they will continue to exist independently in the form of energy. When a substance like water is destroyed in our atmosphere, the elements constituting it would enter other substances such as air, earth or fire.

Although we may observe superficially that water, upon destruction, disappears from our atmosphere, the abstract elements continue to exist in other substances. It is only possible to destroy the material combination (the substance taking shape and size as water) while it is impossible to destroy the basic abstract elements constituting the substance of water itself. This applies to all other substances as well.

In the same way that abstract elements cannot be destroyed, it is impossible to cause the formation of any new units of abstract elements. We cannot, for instance, make water from absolutely nothing. If we were to make water, we must, in the first instance, have the basic abstract elements essential to make water. If we assemble the correct proportions of the units of abstract water, abstract air, abstract heat and the abstract earth, the substance we refer to as water could be produced. Without this process, it would be impossible to produce the substance of water. In other words, we may weave the four abstract elements together and produce any desired fabric but without the aid of the fibres of the four great abstract elements, neither the threads nor the fabrics we desire could ever be made. We can only weave the threads from the existing fibres, but we cannot ourselves produce these fibres, nor can we destroy them.

When we assemble a fabric of a material object with the aid of the fibres which exist in the form of the four great abstract elements, we may certainly claim the fabric to be our creation and claim ownership of the fabric. We cannot, however, claim that the fibres with which we made the fabric from are also our creation. Since we cannot lay claim to the production of fibres, we can only take credit for assembling the fabric from fibres. The four great abstract elements could be assembled and kept bound together for a considerable period of time in the form of a substance. While additional units of abstract elements cannot be made, the units of four great abstract elements, however, circulate and often form one kind of substance at one point and another kind of substance at another point. They never remain in any of the substance permanently.

All material objects that may be produced, all material substances that may be formed and the entire universe itself are, sooner or later, subject to destruction. The destruction of a material object causes only the destruction of the combinations of substances in the material object; it does not cause the destruction of the abstract elements constituting it. Similarly, the destruction of the material universe causes only the destruction of the formation of the material universe; it does not cause the destruction of the abstract elements constituting it.

In the same manner, the destruction of the human bodily form does not equate to the destruction of the abstract elements constituting the human bodily form. It is the same with other beings. The destruction of each material form merely means the movement of the abstract elements settling in some other combination or existing in their independent and individual forms. The total destruction of the material universe, therefore, can never occur.

The present material universe is bound to be destroyed but the abstract elements constituting it will never be destroyed so that, after a long lapse of time, the same units of abstract elements would gather together and cause the formation of an entirely new material universe. The different material universes arise by themselves in space, but it is the same units of abstract elements which, going on forever and ever, make them. These abstract elements never cease to exist and therefore, the basic material substance for the formation of new universes remains absolutely undestroyed. When each material universe collapses, the units of abstract elements remain in the form of free energy distributed evenly over space.

We must remember that although the behaviour of the units of four great abstract elements is demonstrable in the modern science laboratories, it is impossible to destroy any abstract elements within the same domain. All that is done and could be done in any experimental laboratory is to process the different abstract elements and thus cause them to display their intrinsic qualities more and more intensely.

Just as rain is caused by a process of circulation of the four great abstract elements, the travel of heat is also a process of circulation. The sun is not producing any fresh supplies of heat. All the heat that the sun holds is what it has gathered together from its environment. Just as the oceans gather more water from rain and rivers whilst submitting themselves to the process of evaporation, the sun also gathers into it more heat from the neighbouring planets and stars. All the heat that emanates from the sun onto this earth does not accumulate on earth itself. If that were to happen, this earth within a short period of time would turn into a blazing ball of fire. The fact that this earth maintains a constant temperature indicates that the heat it receives from the sun is radiated away either to other planets or back to the sun itself.

All units of abstract elements are subject to circulation but during the course of their circulation even between planets and stars, they neither produce nor destroy more units of abstract elements. Material objects can be constructed and with the ability to handle the requisite forces being available to us, it is not impossible for us to construct an entirely new planet or an entirely new star. By virtue of having been constructed, they can also be destroyed.

The mind, however, is an infinite phenomenon and since the mind is not discernible to the environmental sensory organs, it is difficult to understand its behaviour. Like all other infinite phenomenon, the mind has no beginning and therefore it has no ending. The units of mind come from infinity just like the units of the four great abstract elements and the sky element, and the units of minds are proceeding towards infinity also like the five infinite elements. Just as the four great abstract elements circulate from substance to substance, the mind also circulates from life to life. Just as the four great abstract elements cannot remain in any formations of substances perpetually, the mind too, cannot remain in any formations of substances perpetually. The mind is a force more powerful than any of the four great abstract elements and even the weakest unit of mind is several times more powerful than the strongest unit of abstract element. This bring the mind to a position of supremacy amongst all the infinite units of existence in the universe. The mind maintains this supremacy and the forces of the units of mind keep all the units of elements in control.

The difference between the units of mind and any of the units of matter is that the units of mind are subject to variations in their strength, whereas the strength of the units of elements are constant. A million units of mind would not produce a force exactly one million times greater than any one particular unit of mind. The reason is that each unit of mind has a different strength and the addition of two units of mind would add together two different forces of mind. It often happens that the force of a single strong unit of mind is more powerful than a million weak units of mind. The force of the mind of a human being is greater than the aggregate force of trillions of bacterial units of mind.

By adding many units of abstract element of heat to a substance, the substance becomes hot. However, we cannot add to the strength of minds together simply by multiplication because of the different strength in each unit of mind. Therein lies the main difference between the units of abstract elements and the unit of mind. The process of material life consists of a merger of one or more units of abstract elements with a unit of mind. We must remember that in the merger, the different units do not fuse together completely and that the individuality of the various units is always preserved. In the formation of a material life system, the unit of mind is more active than the units of abstract elements and consequently the unit of mind brings to bear more force.

As mentioned earlier, the tendency of the unit of mind is to gather into its hold as many units of abstract elements as it has the strength to control. Therefore, if the unit of mind is powerful, it would gather into its system a large number of units of abstract elements. When the limit is reached, the unit of the mind ceases to take in additional units of abstract elements.

During the Big Bang, depending on the strength of the minds, all beings from the lower category of elementary life such as viruses, bacteria, insects, animals, marine life together with humans, start to form simultaneously. The model of explanation put forward in this paper is not in agreement with the theory of evolution. It is during the so-called moment of creation, that all animate and inanimate matters were formed. Due to the speed of the mind, all beings formed at the same time, together with mountains, trees, oceans, planets and stars including all the inanimate matters in the universes. Even amongst the most elementary life, beings developing during this time are born without the support of parents. Beings are born without conception. We can argue that this is the origin of life and the universe. However, life and the universes will collapse after a period of time. If we assume that there is an infinite number of Big Bang, and that the four great abstract elements are infinite by nature, there is actually no origin of life.

Based on what was discussed above, at the moment of creation, living things were formed without conception and without the support of parents. With regards to the chicken and egg argument, this would make the chicken first. Due to the limitations of human mind to comprehend a way to understand life, universe and the moment of creation, the concept of God is a convenient alternative. However, the concept of god leaves plenty of questions, especially as humans grow more intelligent by the day. The explanation about the moment of creation where living things were formed without parental support could be helpful. Subsequent to the moment of creation, living things require parental support except for those in the lower categories as discussed earlier. This contradicts with the one cell theory where all animals, human and plants originate from a single cell.

When A Human Dies

The death of a material being involves the separation of its unit of mind from its material body. In the case of natural death due to old age, the unit of mind abandons its influence on the physical system in orderly and gradual stages. They will first lose the senses with regard to the eyes, ear, nose, tongue, and touch in that order. The last organ to go out of action is the heart and that occurs only at the time of death. Even in the case of an abrupt death, the sequence remains the same.

At the moment of death, all the forces of the unit of mind, which are spread throughout the physical system, concentrate in the heart and a strong wave of vision flashes out. The maximum strength of the evolution current of the mind gets focused in this flash of vision. At the moment of death, the mind ceases to utilise its energy to maintain the physical system. The energy will concentrate in the heart and flashes out in a vision of death thought. Ignorance and greed bring our thoughts back to the cycle of life. The vision of death will halt at the highest possible plane of existence once greed is produced by ignorance. The exit vision will rest on a new plane of existence depending on the strength in the evolution current of the mind.

Thus, the old material system dies and a new material system is born. The death vision is the last beat of mind vibrating in a material system of a living thing. Some people believe that they will head to heaven or hell. Human's attachment to the self is so strong that they wish to have continuity even after life. By this model, a human life is a process. It starts at birth, exists in the process, expires at death and ends. Then, the cycle continues in perpetuity.

Though the mind will travel and migrate, the human never actually never dies. This is because the mind is made up of the 4 abstract elements which is unformed, undestroyed and exists in its infinite nature. It just migrates from one physical body to another depending on the strength in the evolution current of the mind. The time it takes to leave the old system and migrate to a new system is instantaneous. Distance is not a barrier to the migration of minds. To this understanding, life on other planets is possible.

When the last beat of mind migrates to a new field, conception take place. At the first beat of the mind in the new field, consciousness occurs and starts to gather elements to cover itself, thus the mental and physical development of the new being begins to take place. During this stage of the development, sensory organ are formed.

Afterlife

The Greek philosopher Plato contemplated on the afterlife. The understanding of afterlife helps to remove the notion of identities. It helps to provide a full and complete understanding of life, universe, the moment of creation and the continuity of the universes together with the infinite number of Big Bangs. Afterlife refers to the migration of mind from the old material system to a new material system.

Law of Condition

It is important to realise the laws of nature. It helps to understand why sometimes God does not answer our prayers. It is not correct to say things happen for a reason. As mentioned earlier, what is more appropriate is when all the required conditions are met, things will happen. This is one of the laws of nature which is the Law of Condition.

In our daily life, we are met with pleasant and unpleasant happenings. These happenings occur because the conditions were right for it to happen, so it happens. While we cannot know what the past causes were, it is advisable to keep enough positive energy in the mind. One of the prerequisite conditions for things to happen is the condition of the deposit stored in the evolution current of the mind.

God helps those who help themselves. It is a situation where we are trying to create the conditions for the outcome we pray for. Saying “God has his own way of doing things” comes about when our prayers don't come true. In truth, it is the Law of Condition that governs.

Law Of Process

Everything in the universe is in a process. Everything in nature is in motion including historical figures, worldly issues, monuments, historical sites, etc. With regards to abstract elements in living things and non-living things, they are all in motion. Everything that begins will first go through the stage of birth, being and end. Everything is in process.

Law Of Cause And Effect

Every action is followed by reaction. Every effect is a result of the cause. But for the effect to come about, all the conditions must be met, which include the condition of the deposits in the evolution current.

Sometimes a person can commit many wrongdoings and yet continue to enjoy a pleasant life. This is because the positive deposits in the evolution current of that person is in abundance. While his wrongdoings do burn out the positive potential, its abundance means he will continue to experience enjoyment of life. Negative potential will begin surfacing when the positive potential becomes insufficient. When all the conditions are met, the effect will come to pass.

Blindness Of The Third Kind

Blindness of the first kind refers to physical blindness, those who are unable to see since birth or lost the ability from some tragedy.

Blindness of the second kind refers to the state of mind when we fail to see a real-life situation for what it really is. In our daily life, when defiling thoughts like jealousy and anger are so intense, we lose sight of the actual situation.

Ignorance of life and the universe cause blindness of the third kind.

The recent suicide deaths of Robin Williams, Anthony Bourdain and Kate Spate were very likely caused by fear, anxiety and maybe depression. These unfortunate deaths show they had blindness of the third kind and likely suffered severe last minute blindness of the second kind. While they were in the state of blindness, they did not realise that they could not see.

Do Humans Have Free Will?

In one of the talks by Dr Ravi Zacharias titled 'Atheist Scientist Challenges Christian Apologist', he was asked by a scientist if humans have free will or is it all predetermined by god as per the Bible?

Doctor Zacharias quoted John Polkinghorne, a world leading quantum physicist, and also another world leading physicist David Berlinski, who wrote the Devil's Delusion. In addition, he quoted the opinion of Stephen Hawking, who had taken back his initial opinion that there is no God. His posthumous book "Brief Answers for the Big Questions', however, reiterated his opinion that there is no God.

The present is the point that separates the past and future. While the past is history, the future is uncertain. In discussing the topic on whether a human has free will in his life, it is necessary to separate the interval before and after conception. From conception to the moment a human has the ability to decide and from the moment a human has the ability to decide until the death moment of his life.

Apart from the assumption that the afterlife is a natural process, some of the laws of nature such as Law of Cause and Effect and the Law of Conditions must be clearly understood. A complete correct understanding of apparent truth and ultimate truth is equally important in this discussion too.

Before conception, the mind with its sufficient positive deposit to be born into the category of human, will settle into a family, society and country which is suitable for the mind to express the potential stored in the evolution of the mind. The potential of this mind is dependent on the past actions that has been committed. In this aspect, the mind has no free will.

Humans have been born into this environment because of the deposits in the mind to allow for the free flow of the reactions which has matured and fructified at the time of conception. In this aspect, human has no free will because the past cannot be changed.

After conception to the time that a human develops the ability to decide, they have no free will. After a human develops the ability to decide, he or she will have free will until the very last moment of their lives.

The potential in the mind in allowing the free flow of the reactions will lay out the possible journey mapping the deposit in the mind. We can imagine this possible journey similar to the river of life. Same as the river that flows only in one direction, humans can only grow old while boulders at each cross section of the river of life represents the obstacles that we experience in our lifetime.

The future is uncertain because we humans do not know our past actions, hence we do not know the entire profile of all the cross sections of the river of life ahead of us.

A Canadian fashion designer working in New York decided to die in style after she was diagnosed with cancer and was given a few months to live by the doctor. She travelled to either Tibet or somewhere near Tibet and started living a monastic life.

One day after telling her story to a monk, she was told that since life is so important to a layman like her, maybe she can try to increase the energy in her mind in order to create a condition where she does not need to follow the path of dying from cancer.
She moved back to her fishing village in Canada. There, she would buy a quantity of shrimp from fishermen and release them to the sea on a regular basis. She continuously did that until her health condition improved to the extent that she was under remission. With a healthy body, she went back to the fashion industry in New York but under stressful conditions, she went into relapse. Repeating what she has done before by saving lives to increase the

energy in her mind, she has given up on the stressful lifestyle and is alive as far as the author knows.

This is an example of the possibility to alter the profile of the cross sections of the river of life. A situation where the increased energy in her mind is so tremendous that it provided the conditions, so she did not need follow the path of dying from cancer.

The profile of the cross sections of her river of life could be that she has to die from cancer. It could well be after saving the lives of the shrimp, she will still die from cancer.

On the question whether she could still die from cancer after all the effort to increase the energy in her mind, we need to touch on apparent truth and ultimate truth. Apparent truth is subject to ultimate truth, which again, is in accordance with the law of conditions.

Some humans are born into countries where euthanasia is practiced. The apparent truth is that the patient is subject to the laws of the countries. But the result, whether a human has the free will to die is subject to the ultimate truth.

The result could well be that the river of life was mapped out to accommodate the deposits in that human's mind. In this instance, where a patient lives in a country where euthanasia is not practiced, he has no free will to die. This is only the apparent truth where he submits to the laws of the country he lives in.

But the final situation is a result of the ultimate truth. Yes, subject to ultimate truth, human has free will.

Just like a burning candle can go off in three different scenarios, the wick has run out, the wax has run out or a strong wind came blowing. Humans can die of natural death following the river of life or meeting an accident and died prematurely.

A 21-year-old, using his free will to decide to take drugs and speed in a vehicle, was actually creating and providing a condition for accident to happen to take his life. It could well be that his profile of river of life was such that it lasted and flow for 21 years only. It could also be that his profile of river of life should be some 70 to 80 years, but he used his free will to create the conditions for the unpleasant to happen.

Humans do not know the profile of the river of life which was mapped out based on the potential in the mind during conception, but human has free will.

Nobody knows the profile of the river of life for Anthony Bourdain, Kate Spate and Robin William. They could have died prematurely or followed the length of their river of life. But they had their free will.

In one of the talk where he was asked the question 'Does suicide send you to hell?', Dr Zavi Zacharias responded that it is an ultimate disrespect to god, a violation of the image of god and if it was him, after taking his own life, he will not want to meet god. He advised people not to commit suicide. He did not answer the question directly.

Apparent Truth Versus Ultimate Truth

Dr Ravi Zacharias once said in his speech, truth is more powerful than a nuclear weapon. But we have to define truth by separating the apparent truth from the ultimate truth.

When a person works hard and smart, is always at the right place at the right time meeting the right people and is financially comfortable, it is only an apparent truth. Another person could be equally smart and go through similar opportunities yet fail to make it. Both the examples are subject to the Law of Conditions.

By nature, human minds are more inclined to defiling thoughts which leads to evil actions. Pope Francis, on 25th Feb 2019, made a statement that the abusive clergy who committed sexual abuse are tools of Satan. Actually, when there is no god, there is no Satan. Again, sexual desire is a form of defiling thoughts and human by nature are more inclined to defiling thoughts. Due to this natural pull of human mind towards evil thoughts and actions, there is a common debate on why God allows evil to exist in this world.

Only Those Recognised By Nature Can Be Considered As The Ultimate Truth

In order to distinguish the ultimate truth from apparent truth, position yourself as nature looking at the world from afar. When the sun shines on one part of the earth, the sun does not recognise borders of countries, nationalities, race or religions. The way nature makes its recognition is through the ultimate truth. The way events take place, things that happen to a human based on the mapping of the river of life, one's actions and reactions and the law of conditions are all part of the ultimate truth.

Observation 16 - Fate Is Only True For The Past

Everything happens when all the conditions are there for something to happen. Everything that has happened in the past, we can if we so choose to say, is fated. However, fate is only true for past events. The future is uncertain. The profile of the river of life in the mind waits for the appropriate situation to fructify all the potential in the evolution current of the mind. Our present life is not completely predetermined. Our past actions influence the flow of our lives, directing them towards pleasant or unpleasant experiences. However, our present actions towards the present situation has effect on the predetermined flow of life. Depending on the intensity of our present actions and the predetermined cross section of the river of life in the mapping drawn up in the evolution current, humans have the ability to decide their course of action. Hence, fate is only true for the past and future is uncertain.

Tsunami-prone Areas

We are born in different countries because the character in our evolution current required such environments to allow for the free flow of the reactions which had taken their places to mature and fructify at the time we took conception. The particular area, society or country in which we are born in corresponds with the mature impressions which are about to fructify in this life.

A 9-month-old baby is born in a tsunami-prone area and is killed by the natural tragedy. The baby has not acquired the ability to decide and take actions in his or her life to enhance the positive deposits in their evolution current so as to avoid the tragedy.

In this instance, it is due to the deficiency in the positive deposit in the evolution in the mind of the baby, which was mapped out according to the river of life. For those who believe in fate, this is what they would label as 'fated'. For the believers in God, they may have some difficulties to explain why this unfortunate tragedy happens.

Observation 9 - Good News And Bad News Are Just Information

With reference to the Law of Condition, Law of Cause and Effect and the Law of Process, everything is just the way it is. There is no good and no bad. Similar to how both music and noise are simply sounds. When something pleasant happens due to the Law of Condition, it happens. When the unpleasant happens, it is because the conditions are right for the cause to fructify. So, if you put all the three laws of nature together, there is no good and bad. It is just the way it is.

Observation 10 - Both Music And Noise Are Simply Sounds

Music and noise are apparent truths. Ultimately, they are sounds. Rock music may be appreciated by some while others prefer the opera. One may categorise the other's taste as unpleasant noise. That is how we live life wanting music and hating noise. Little do we pay attention to the ultimate truth that both are sounds. Similarly, we only wish to have the pleasant and avoid the unpleasant. Little do we know that it is just the way it is. If human can live lives by seeing the ultimate truth, living would be less stressful.

Observation 12 - Not Half Empty, Not Half Full. It's Half A Glass

In our daily lives, we often hear the argument between viewing situations as glass half empty or glass half full. This argument is made to differentiate those who are supposedly optimistic or pessimistic. A person who chooses to see things as glass half full, may think its half empty at another time due to the conditions of the mind. The mind is a force and it is subject to variations in strength at different periods of time. The ultimate truth always holds true, no matter what the conditions of the mind and no matter when.

Deposits In The Evolution Current Operate Like A Bank Current Account

When the deposits in the evolution current of the mind is sufficient to attract the pleasant, we will enjoy pleasant moments. As we enjoy a pleasant moment, we are burning out the positive potential that has been stored in the evolution current. During the pleasant moments, the wise will continue to do wholesome actions to replenish the positive potential while avoiding negative thoughts and actions.

Appearance of negative thoughts and actions reduce present and future pleasures. The present sensations of pleasure and pain are the reflections of reactions of our past causes and it consumes the growth in the evolution current so as to reduce the stock in the deposits. Imagine sowing seeds while reaping the harvest. This will result in a never-ending procession of periods of pleasure.

When a person is not contented while in pleasure, negative forces cut off the periods of pleasure abruptly and invites the fruition of negative potential. When a person suffers, negative forces are extremely strong even with the slightest provocation. While in pain, a human should practise positive thoughts and actions so as to reduce the duration of suffering and to accumulate positive potential.

A delay in the fruition of positive and negative potentials accumulates the strength of fruitions. Thoughts of action are those that inspire us to carry out the action. Pleasure and pain are not thoughts of action. Pleasure and pain are the passing reactions.

The Chinese has a saying that says ‘wind and water take its turn’ which simply means when one enjoys the fruition of purity seeds, after sometime when the purity seeds are depleted, it will be followed by defilement seeds. Thought causes impressions on the mind. Deep impression causes actions. Adding additional positive potential can cause past positive potential to materialise.

Hunting for pleasure is an act to kill beings which can result in a huge deficit of positive deposit in the evolution current. When an animal kills another animal for food, the purpose is for survival and not for pleasure. The advantage of being an animal is that when they kill for food, they are ignorant of their act. Hence, there is no bad intention and as such, there is no depletion of positive deposits in the evolution current. With this, animals can slowly climb the ladder from a lower category to a higher category when their mind migrates. This is because, as an animal, they suffer which diminishes their negative potential. At the same time, while killing for food, they do not diminish the positive deposits in their evolution current.

Similarly, the positive deposits of an ignorant man in doing negative actions is reduced but not as much when compared to a developed human. A developed human is considered as someone who knows and is conscious of the consequence of defiling actions. Mindfulness and having a developed mind are responsibilities. It cuts both ways. In this particular case, ignorance is bliss. All types of conscious destruction of life causes negative potential. The higher the category of living things being destroyed, the graver the reduction of positive potential in the evolution current.

Luck

Luck is a common phenomenon in our daily life. There is no arithmetic or formula to calculate the timing of when will we be lucky or unlucky. It could be possible to equate luck with the balance of deposit in our evolution current. It is possible, to some extent, to increase the possibility of getting lucky by increasing the amount of positive deposit entering the evolution current from certain kinds of activities. By the explanation of this model, humans can, subject to the law of conditions, create luck.

Observation 5 – Alzheimer's Cannot Be Cured, Could Be Delayed

Briefly, as explained in medical science, Alzheimer's disease is caused by the progressive degeneration of neuron population in a particular area of the brain. The areas of brain involved are in the enthorhinal cortex, the hippocampus, the high-order associations cortices of the temporal, frontal and parietal regions. The neuronal damage and the attending loss of synaptic damage and the attending loss of synaptic density disables several neuronal networks essential to learning and retrieval of memories.

In addition to neuronal population loss, damage to 'cholinergic neurons' results in a loss of delivery of brain chemical called acetycholine to the cerebral cortex. Clinically, Alzheimer's disease is characterised by relentless impairment in memory and decision-making that generally begins slowly and worsens over a decade or so. They fail to learn new events. They will also lose recognition of familiar faces, places or situations. Poor judgements and deterioration of social relationships commonly become worse.

Once someone gets Alzheimer's disease, it is always a downhill decent. Fear of getting Alzheimer's disease is one of the harmful causes of the disease. This resembles the computer game called Tetris. In the game, the objective is to avoid the blocks from building up to the top. Naturally, one begins to feel anxious as the lines begin to fill up and a sensation of panic sets in. This further worsens the situation, similar to how worrying about Alzheimer's might actually cause it to happen.

Present medical science has neither found a cure nor how to prevent Alzheimer's disease. Humans cannot go against nature by not growing old or by not losing the required amount of neuron population as we age. However, we can try to leave those records or thought of impression in our minds untainted so that even with a less than healthy neuron population, it is sufficient for the retrieval of records, or memories.

Thoughts of defilements will form obstruction for the retrieval process as compared to records of pure thoughts of impression in the mind. A weak torchlight can still shine on to the records beneath if there are less obstructions. It can still penetrate even with weaker energy in store if most of the records of thoughts of impression are clear. This method may help to delay the disease and if we are able to delay the disease until the last days of our lives, we would have prevented the disease.

Western doctors have, in fact, found that the state of mind has a bearing on the physical health of a person. Diseases are sometimes caused by a stressful mind. Prolonged depression, anxiety, fear or hatred can lead to chemical imbalances and other disorders which can cause diseases. Unhealthy mental states may lower our resistance against infection. Other studies showed that anger or agitation causes the production of the chemical called epinephrine, which increases blood pressure, heartbeat and oxygen consumption.

Fear and anxiety cause the secretion of a greater than normal the amount of acid, which attacks the intestinal membranes, resulting in ulcers. A prolonged state of nervous tension is known to impair the automatic nervous system in such a way that it affects the digestive function and elimination of waste. This results in constipation and self-poisoning. Some of the negative formation on the mind affects specific body organs: grief to the lungs, hatred to the liver and fear to the kidney.

Possible Reason For LGBT (Lesbian, Gay, Bisexual, and Transgender)

The mind of a human being has 28 elements, of which 2 are sex elements. The predominant sex will prevail while the other sex which is dormant will slowly gather strength. A male human will be using his male sex element throughout his lifetime. When the process of migration of the mind from one physical body to another takes place, if the male sex element is stronger, then the human will live the life of a male being. On this occasion, the human is living a full life and experiences death due to old age. However, as mentioned earlier, there are instances where accidents occur, when that male human has yet to fully utilise his male sex element. When the conditions are right and this mind reaches maturity to occupy a new material system of a living thing, the difference in strength in the male sex element and the female sex element is not substantial. Hence, the complication of sexual preference and even the sexual organs.

Gun Control

Humans are more inclined to blindness of the second kind. Knowing how a human mind works, it is advisable to have gun control where only enforcement officials has the right to carry guns. Even then, the possibility of enforcement officials suffering from mental disorder and committing a crime is always there. However, the probability is lower seeing that the number of gun-carrying humans would be far less on the streets. Also, background checks are more feasible given the fact that the number of enforcement officials are less compared to the general public.

In recent years, the number of school and college shootings in the United States have not been decreasing. There was a suggestion to increase the age limit to 21 for those who can legally purchase firearms. This proposal will not work, referencing to the case where a senior man aged 64 opened fire on a shooting spree in Las Vegas in 2017. Age is not the source of problem.

As recent as February 2019, another shooting in Aurora, Illinois took another 5 unnecessary lives and injured dozens. We have become desensitised to these regular shootings. As long as the public have the right to carry firearms, and since public are humans, the shootings will never end. If world leaders realise the nature of human mind, there will be no guns in the hands of the public.

28-year-old Brenton Tarrant, the person responsible for the shooting in Christchurch, New Zealand, actually had a 87-page manifesto. He must have prepared this a while ago and as the mind gets clouded with blindness of the second kind, the manifestation of defilements in the mind gets stronger. One of the newsreaders stated that his mind must have fermented with evil thoughts for a long time. Finally, he committed the shootings. Some 50 unnecessary deaths, a result of the identities of race, religion and nationalities.

Observation 51 - We Board A Plane Because We Assume The Pilot Wishes to Stay Alive

Human minds are more inclined to a negative thought's natural pull. A depressed mind is not noticeable when we look at a pilot. A pilot is susceptible to nervous breakdowns, anxiety and fear just as any other humans.

Observation 52 - One Who Has Never Woken Up Does Not Know The Difference Between Being Awake And Asleep

When the turtle goes back to the sea and tells the fishes that there are animals who can live above water and walk on the ground, the fishes may find it difficult to understand and comprehend.

Observation 50 - It Is Dangerous If A Blind Person Does Not Know He Cannot See

If a blind man knows he cannot see, he will carefully move about without hurting himself and others around him. The situation is very chaotic when he does not know that he cannot see. The situation is made even worse when a blind person leads another blind person. In our daily lives, a majority of us suffer from blindness of the second kind.

Something That We Do Not Want To Lose May Not Be Something That We Need

Greed and attachment in human always results in what we want rather than what we need.

Observation 40 - Without A Mirror, You Can Never See Your Own Eyes

When something goes wrong in our daily lives, we tend to look for someone to blame or even blame the situation itself. By nature, our eyes only see outwards. Similarly, by nature, humans suffer from blindness of the second kind (depression, anxiety) which, to a great extent, is due to the blindness of the third kind (a lack of the complete understanding of life, universe and the mind). When a person is having blindness of the second and third kind, one tends to see pain as pleasure and pleasure as pain.

Observation 53 - Confidence Is Developed Not Practised

When we tell our children to be confident in the examination hall, we are actually asking them to practise being confident. If our child has not studied hard before the exam, practising confidence would not help. However, if our child has been consistently studying hard with guidance from parents or teachers, their confidence will develop. Similarly, going onstage to give a speech or a performance. With sufficient practice, confidence can be developed. True confidence cannot be practiced.

Magic

There are few fundamentals in magic. Practice, precision, speed, technological advancements, preparation in terms of getting the right instrument and performers. Though we are not able to explain all the preparations and arrangement for all the magic that is successfully performed, we do know all magic follows the above-mentioned fundamentals. Similarly, this model of explanation on life and the universe may not be able to cover all the details, but it may serve as a good perimeter in understanding life and the universe.

Does Artificial Intelligence Have The Ability To Imagine ?

Robotic advancement and new technology are helpful to humans. However, the robot has no mind and lacks the human ability to emote such as having compassion, kindness, jealousy, greed, confidence and self-pity. Robots can never replace human. Robots can store data and memories but they are not able to have emotions associated with the mental formations such as they do in the human mind. Humans are subjected to the afterlife because the human mind migrates while the robot does not have one.

On the 12th of February 2019, CNN reported that a computer lost to a human in a debate. Generally, the ability to debate well depends on the presentation and the ability to articulate facts based on reasoning. All the reasoning ability acquired by the computer are the data and memory input that has been inserted into it. At best, computers are able to debate based on these data. And as far as the computer is concerned, they are facts.

The intelligence that is derived from these inputs are artificial intelligence. Human intelligence can go beyond knowledge. Humans have the ability to argue and manipulate facts to their advantage. Humans can bring the computer to a platform that the computer could be at a disadvantage at in terms of weak data collection.

It will be a big achievement in modern science if a robot can be installed with a feature to monitor the formations of its mind. When the indicator shows an intense negative formation, obviously suffering from the blindness of the second kind, the robot will know to be careful in making the decision. If it can be installed in a human, an electronic indicator of the similar function, it will certainly be useful for humans to carry out their daily lives.

Observation 55 - We Are All Prisoners Of Our Own Minds

We are all prisoners of our mind. To keep our body fit, we exercise. To keep our mind fit, we should try to keep the mind stay still. Constantly, at every waking moment, conscious and unconscious thoughts arise in our mind. If we practise meditation and mindfulness, we will have the ability to notice the thoughts, to analyse the nature of the thoughts and digest over it. Once the thoughts have gone through the filtration, only then we act. If we notice that those thoughts are unwholesome, we should just notice them without any action or reaction, but mere observation. Dissolving thoughts is a form of suppression. Between suppression and expression, there is observation.

Most of the problems stem from the fact that we begin dealing with other people without first managing our own minds. It is wise to realise that when our minds are clouded in darkness, they abhor light.

The basic problem of life is that it is unsatisfactory in nature. Things happen that we do not want, things that we want do not happen. We are ignorant of how or why this process works, just as we are ignorant of our own beginning and end.

Our personal dissatisfactions do not remain limited to ourselves. Instead, we keep involving people around us with our ways of handling ourselves. The atmosphere around each unhappy person becomes charged with agitation, so much so that all who enter that atmosphere may also feel the agitated environment. In this way, individual tensions combine to create the tense environment at home, at the work place and in the society.

When we wake up in the morning, take notice of what's in the mind. Chances are that it is empty for about 2 to 3 seconds. During these moments, the mind has not been filled with thoughts. The mind is at ease. It is this state of mind that people who meditate attempt to achieve.

However, by nature, after the first few seconds, the mind starts to fill with thoughts, arising from the most recent issues, the most impending matters and matters that affect you the most recently. The files start to pile up. One file overlapping the other. Carelessly, we allow the manifestation of the thoughts and the mind starts to get clouded.

Before being able to handle our minds, we have to start dealing with people around us. With a clouded mind, we begin dealing with the day's issues. That's where the problems begin. However, if one realises the nature of the mind, watches the mind at every waking moment, filtering thoughts while avoiding reaction to defilement thoughts, problems can be avoided.

Our mind works incredibly quick, hence it requires practice to observe the mind. Mindfulness is required at every waking moment. The mind and body are interconnected. In order to keep the body at ease, it is necessary to keep the mind calm. Understanding the thought processes will be helpful to keep the mind in calm state.

Generally, there are four processes involved:

1. Consciousness
2. Perception
3. Sensation
4. Reaction

When a thought arises, the first process, consciousness, is the receiving part of the mind, the act of undifferentiated awareness or recognition. It simply registers the occurrence of any phenomenon, the reception of any input. It notes the raw data of experience without assigning labels or judgements. The second mental process is perception, the act of recognition. It identifies whatever has been noted by the consciousness. It distinguishes, labels and categorises the incoming raw data and makes judgements. This is followed by sensation. So long as the thought is not judged, the sensation remains neutral. But once judgement is attached, the sensation becomes pleasant or unpleasant, depending on the judgement of the thoughts. If the sensation is pleasant, a desire is created to prolong and intensify the experience. If it is an unpleasant sensation, the desire to reject arises. The desire to reject causes tension in the body and the organs. As discussed earlier on the speed of mind, we human fail to take notice of this process. Seeing that life is generally unsatisfactory, tension builds up in our body. Prolonged tension in the body causes unnecessary sickness.

Due to the natural pull of human mind towards negative formations, we are commonly in an unsatisfactory state of mind. The very fact that we were born human is simply because the positive deposits in our evolution current was not sufficient to bring us to the higher category of happy spirits.

As long as we live, we are in a mental jail. We can never get rid of our mind. At best, we can only calm it because our mind was unformed and will never be destroyed. It will migrate from a living thing to another living thing upon expiry of the present biological life.

Observation 54 - Nothing Is Bigger Than Nature

Nature encompasses everything within it.

Observation 20 - Life Is Nothing But Just A Sad Joke

The mystery of the universe and the origin of life are long-standing mysteries. The open secrets to these could be due to the following:

1. Time and space are infinite. Time existed before the Big Bangs.
2. Living things are developed directly from the unit of abstract heat energy and are born without parental support.
3. Reincarnation is a natural process.
4. Gravitational force is directly related to the energy of the minds.
5. The evolution theory by Charles Darwin is inaccurate.
6. The laws of nature are constantly enforced regardless whether the laws are understood or accepted.
7. There is no origin of life.

These open secrets simply lead to one conclusion; life is nothing but just a sad joke.

www.ingramcontent.com/pod-product-compliance
Ingram Content Group UK Ltd.
Pitfield, Milton Keynes, MK11 3LW, UK
UKHW041952190726
13854UKWH00005B/1930

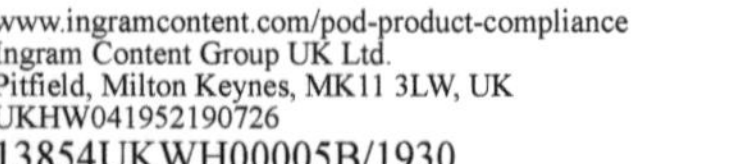

9 781789 556643